INSIDE SCIENCE

AF792456

LIGHTS, CAMERA, ACTION!

Dan Taulapapa McMullin

CONTENTS

THE SCIENCE BEHIND THE MOVIES

Superheroes in action landing on imaginary planets light-years away! It sounds like magic, doesn't it? Scenes like this in movies are only possible because of technological advances made possible by scientific discoveries.

Scientists like Thomas Edison made the discoveries that led to the technology moviemakers now use.

Filmmaking is based on many areas of science and technology, including:

- optics—the study of light
- acoustics—the study of sound
- biomechanics—the study of how living things move
- electronics—the study of electronic devices

For example, developments in the study of electronics have led to the use of computers in filmmaking. When you see a jet fighter fly across your television screen or an explosion in a movie, you are watching images created by art, science, and technology working together.

CREATING IMAGES

The science of imagemaking can be traced back to ancient times. Over 2,000 years ago, Greek thinkers like Aristotle studied how light behaves. About 1,000 years ago, an Iraqi thinker, Alhazen, invented the camera obscura (pronounced *ob-skyur-a*), a tool to capture images. His camera obscura was a tent that allowed light in through a tiny hole. On the wall opposite the hole, an upside-down image of the scene outside appeared.

How a modern camera obscura works

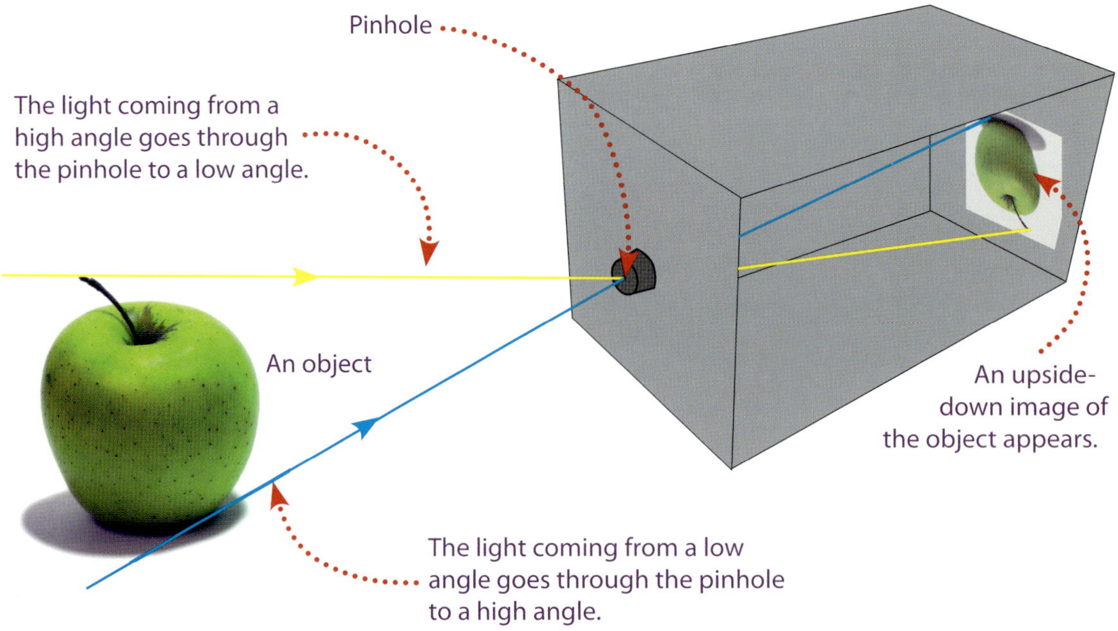

Four hundred years ago, painters started using the camera obscura to **trace** images directly from real life onto paper and canvas.

Make Your Own Camera Obscura

Here's a way to build a camera obscura.

You will need:

- a nail
- thick cardboard
- tape
- a small room with one window

What to do:

1. Ask an adult to help you push the nail through the center of the cardboard.
2. On a sunny day when the sun is not shining directly on the window, tape the cardboard so that it *completely* covers the window and no light can leak in around the edge.
3. Look at the wall opposite. An upside-down image of whatever is outside the window will appear!

Show your friends:

Tell your friends why the image is upside down. Explain that the image on the wall is made by sunlight traveling in a straight line. When the rays of light go through the pinhole, some of them cross. This creates the upside-down image.

In 1802, Thomas Wedgwood and Sir Humphry Davy figured out how to capture the image from a camera obscura without having to trace it. They noticed that some chemicals are photosensitive. Photosensitive chemicals **react** when they are **exposed** to light. The first photographs were called shadow images because they didn't last very long! The chemicals kept reacting until images turned completely black.

Did you know that photosensitive chemicals are found in nature, too? Plants that close their flowers at night use them to detect sunlight.

Scientists solved the problem of how to stop the chemical reaction with fixatives. Fixatives are chemicals that stop the reaction, fixing the image on the paper.

Abraham Lincoln was one of the first presidents to have his photograph taken. Matthew Brady took this photograph of Abraham Lincoln in New York on February 27, 1860.

Moving Pictures

Once scientists had discovered how to capture images, the next step was to make photographs move. After the invention of **flexible**, **transparent** film in the 1870s, Thomas Edison in the United States and the Lumière brothers (pronounced *Loom-ee-air*) in France invented the first modern moving picture cameras and projectors.

At first, movies were silent. What the actors said was printed on the screen. Talking motion pictures, or talkies, followed later, when scientists had figured out how to add sound.

Handle with care!
Between 1895 and 1951, film was often made using a chemical called cellulose nitrate. People must handle cans of nitrate film with great care. They can explode!

Glass Matte

Once the basic technology of motion pictures had been developed, filmmakers used science and technology to do interesting things with it.

A castle, a forest, and mountains painted on sheets of glass

For example, filmmakers who wanted to put a castle in the background of a scene realized that they didn't need a real castle! They painted part of the scene on sheets of glass. Then they shot the film through the glass, with the foreground and the actors in the unpainted areas. This technique is called glass matte.

An actor standing in the real foreground seen through an unpainted part of the glass

Magic Lanterns

Another technique was borrowed from stage magicians. A magician could create a "ghost" by using an image of an actor playing the part of a ghost below the stage. The scene on the stage would take place in front of a large piece of glass set at a 45° angle to the audience. The reflection of the actor combined with the lighting of the scene on the stage created the **illusion** of a ghost. This magic lantern technique is still used by filmmakers—though the technology is now much more advanced.

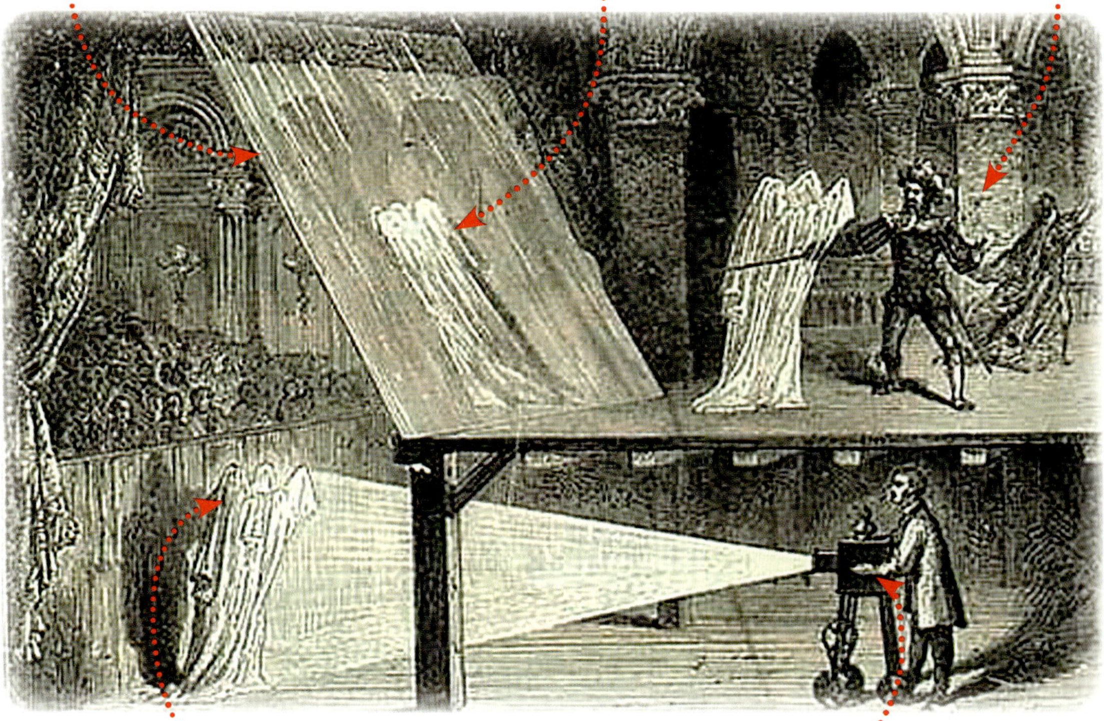

Forced Perspective

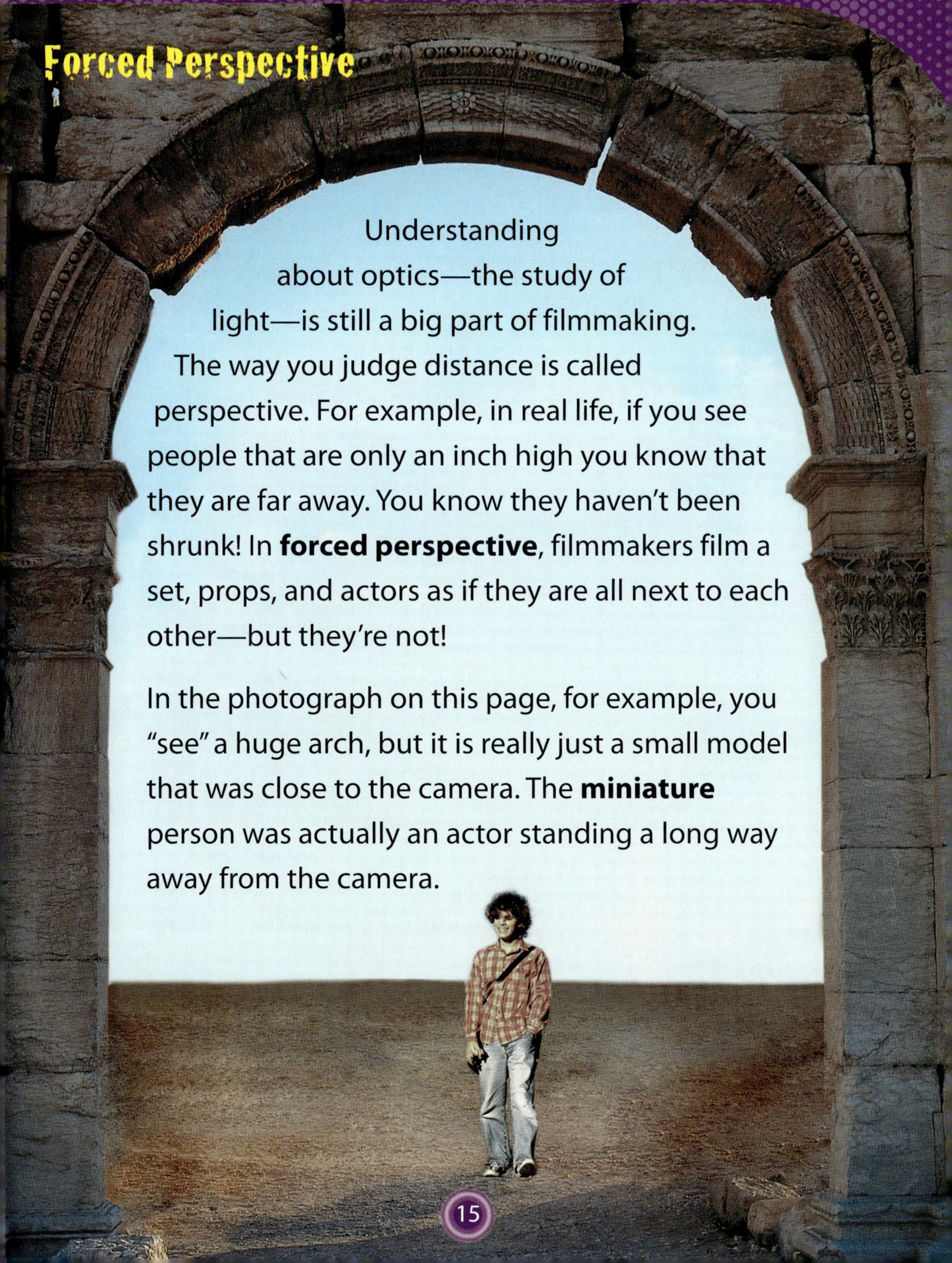

Understanding about optics—the study of light—is still a big part of filmmaking.

The way you judge distance is called perspective. For example, in real life, if you see people that are only an inch high you know that they are far away. You know they haven't been shrunk! In **forced perspective**, filmmakers film a set, props, and actors as if they are all next to each other—but they're not!

In the photograph on this page, for example, you "see" a huge arch, but it is really just a small model that was close to the camera. The **miniature** person was actually an actor standing a long way away from the camera.

Shrink a Friend

How does forced perspective work? Here's a way to see how.

You will need:

- locations
- two friends (you're going to "shrink" one of them)
- a still or video camera that lets you control the **depth of field**
- props

What to do:

1. Choose a suitable location, such as a hill.
2. Position the person to be "shrunk" in the background, up on the hill.
3. Position the other person closer and down lower, in the foreground.
4. Looking through the camera, direct the person in the background to "sit" on the other person's hand. (If one person isn't in focus, adjust the depth of field.)
5. Experiment with different positions, locations, actions, and props.

Show your picture:

You are using forced perspective to trick the audience into thinking that you have shrunk your friend!

Miniatures

Building huge sets for film scenes is expensive—and sometimes impossible. By using miniatures of landscapes, buildings, spaceships, and people, filmmakers can shoot scenes they couldn't shoot otherwise. Miniatures are built to **scale**—everything is in proportion to its actual size in the real world. The bigger the scale of a miniature, the more accurate it will look.

Because real-sized objects move differently from scale models, when miniatures explode in films, the shot is filmed in slow motion.

Minature mothership from the movie *Close Encounters of the Third Kind*

ADDING SOUNDS

Before the introduction of talkies, when movies were silent, a person played a piano in the theater while a movie was shown. This provided the music. In the 1870s, scientists like Thomas Edison discovered how to record sounds.

Recording sounds in the 1890s

Once filmmakers could record sound, they figured out how to **synchronize** sound with the moving picture. "You ain't heard nothin' yet!" were the first words ever spoken in a commercial movie with a soundtrack. An audience heard the actor Al Jolson say these words on October 6, 1927, in New York. After a moment of **stunned** silence, the people in the audience rose to their feet, applauding.

The *Our Gang* movies started in 1922. Sound was added in 1929.

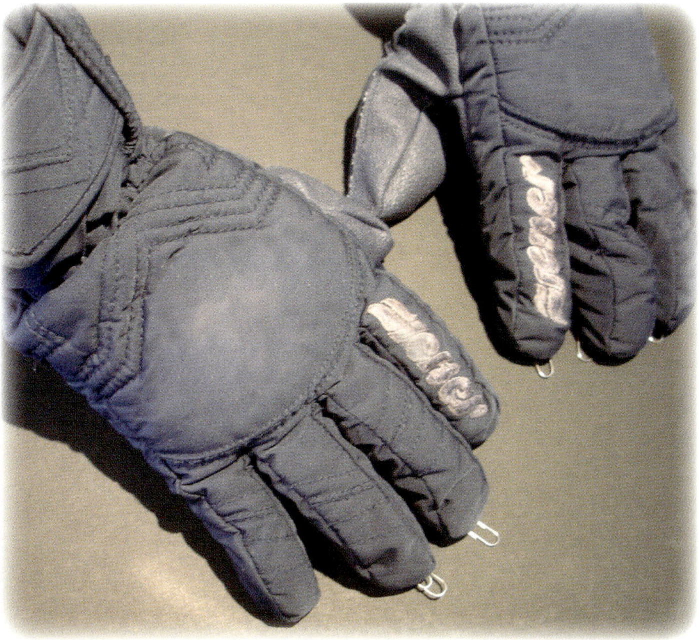

Foley artists use gloves and paper clips to create the sound of a dog walking.

There is an art (and a good deal of science) in getting the sound in movies just right. It's not only a matter of synchronization. **Audio** tracks of different sounds are combined to make film sound believable.

A sound engineer called a Foley artist uses different materials and objects to create sounds to add to scenes. Tricks of the trade include corn starch in a leather bag to make the sound of crunching snow, gloves flapping to make the sound of birds' wings moving, and fresh lettuce squishing to make the sound of bones crushing!

Did you know that some of the sounds in movies are below the range of what people can hear? Instead of hearing them, you feel them! The next time you see something like a train roaring past or a rocket lifting off in a movie, close your eyes. See if you can feel the sound!

FROM COMPUTERS TO VIRTUAL HUMANS

Computers are changing filmmaking. Modern films use powerful computer programs—programs so powerful that they even take into account the slippery way water **molecules** move.

The ocean, the ship, the spray, the fog lights, the waves, and the overall shot of this ship in a storm were all made by a computer. None of this is real!

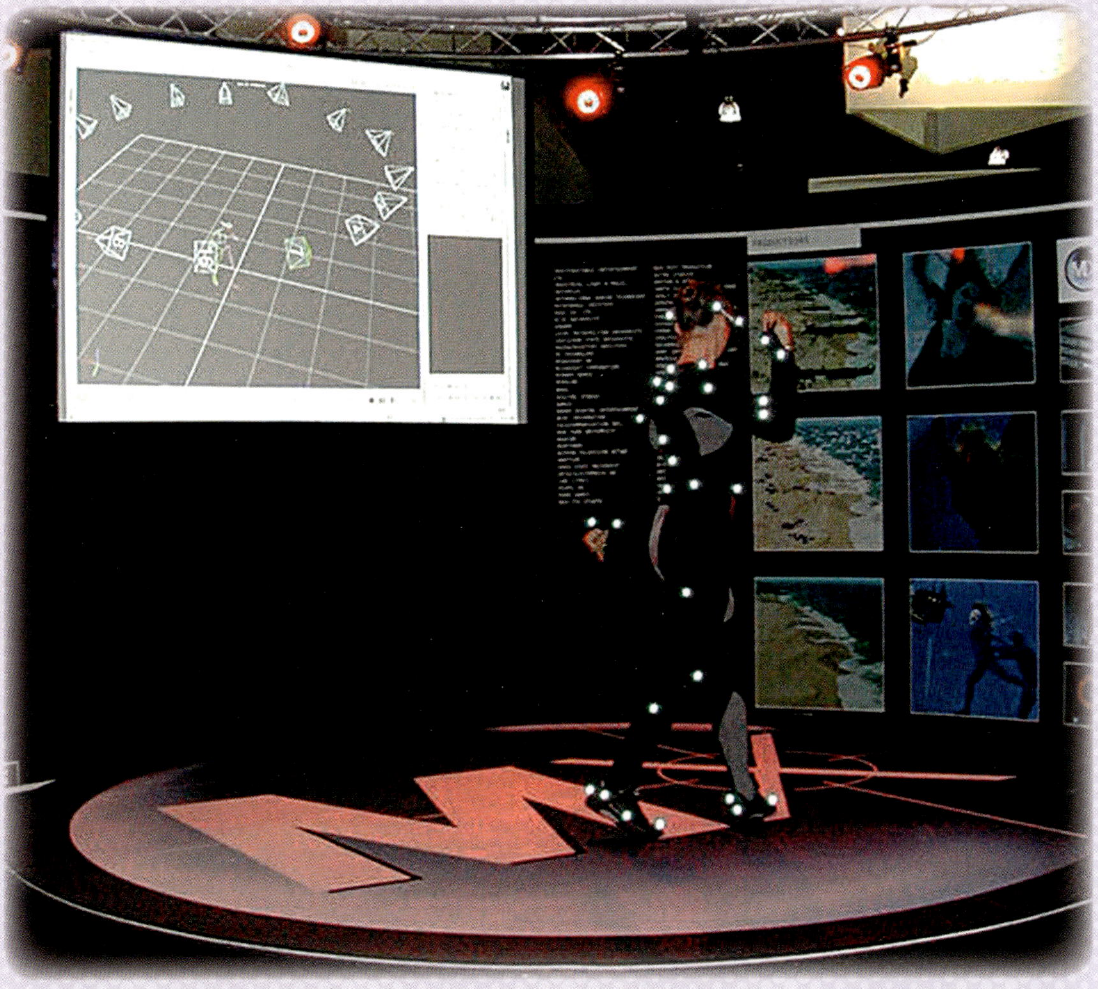

This motion capture suit is hooked up to a computer. The computer scans the reflective markers to capture the actor's movements.

For monsters and other creatures, computer **animators** use sliders to control each part of an actor's face and body, right down to individual hairs and wrinkles. Sliders capture the movements of an actor who is wearing a motion capture suit covered with reflective markers. The computer uses these movements to animate the computer-generated creature.

Computer-generated filmmaking now allows the creation of **virtual** humans. Virtual humans are created by computer engineers and biologists working together. With virtual humans, a filmmaker can make a film without using real actors.

A virtual teacher—through **sensors** built into a computer mouse, she reacts to how you feel while you are learning. The sensors pick up things like changes in your temperature and heart rate. The computer is programmed to respond to this.

From the camera obscura to a virtual teacher, imagemaking has come a long way. Imagine where it might go to next. Is this what a movie might look like in the future? What new scientific discoveries will be needed to bring this about?

GLOSSARY

animators—artists who create images of objects and cartoon figures that move

audio—relating to sound

depth of field—the area that is in focus

exposed—touched by light

flexible—able to bend without breaking

forced perspective—a film technique that makes things look bigger or smaller than they really are

illusion—something that appears to be something that it isn't

miniature—a small model

molecules—the smallest parts of something that still retain its properties

react—chemically change

scale—the proportional size of a model compared with the original

sensors—devices that pick up changes in the surrounding environment

stunned—too surprised to do anything for a moment

synchronize—programming equipment so that it runs at the same time

trace—copy (sometimes just the outline)

transparent—able to let light pass through

virtual—standing in for a real thing